(Context tHe Utterrance = Function : "One")

bLONDe ANTHOLOGY

By
Stanley Alexander MARTIN/Nana baBa jaH-aYe

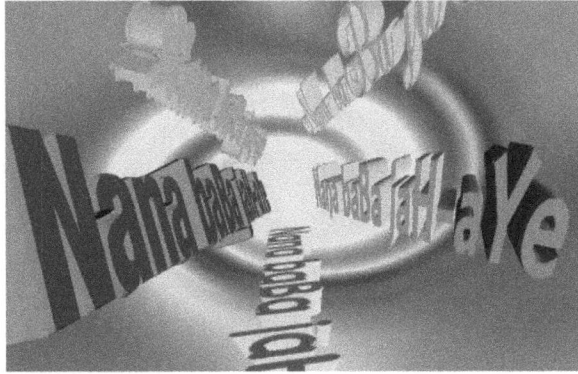

ISBN: 978 – 0 - 9559904 – 0 - 3

Dedicated to aLL….

CONTENTS

One World

"MAMA SKY"

Mama, the Midnight Sky is Dark!
Here, beyond the Colours of the Moon:
Your Horizon is my Mark,
Mama, comes a Holy Tune:
Comes the Sun as Your Lark....
Your Baby Now
Whispers with
the Wish
of Your Heart....

Arrives the Morning Yet?
The Choir singing up the Sun?
Heaven is a Guitar Full of Frets:
Mama, Colour of Leaves are Dun....
Why do We Live Apart?
Why do We Live Apart?
Your Baby now
Wishes to
Whisper with
the Wonders of:
Your Heart....

Noon: the Shy Sun is at a Peak;
Older than Mountains, the Light is the Sing:
My Shadow Now Lies beneath my Feet:
Mama, Inside, This Day Loves Everything:
Evening Must Start....
Your Baby Now
Wishes to
Whisper With
the Wonder of
the Wander of
Your Heart....

The Mountain Stands at a Distance from the Sky:
Here, at This Horizon Line:
Here, at This Horizon Line.....

THE OBSERVER INTERFERES:
-CRITICISM-
Premise

aS i know the Potential of your Id is to know ALL....
yOO, Born of Time/Space: yOOr "Art" translates to a knowledge of All in
the-Universe: the-It.....
Mama Sky
"sHe: All is based on dualism!
But, with Black: White, our Colours point to 8 : 9 Knots in Your String....
But the Horizon Line Points to the fLaw that Dualism Always Suggests 2 : 3
Knots.....
All Waves Must, Too, Point to 2 : 3 Dimensions.....
The Movement of the Sun Points, Too, that We are at Base Two Dimensions with
One
Dimensions of The Eternal Now.....

The Observer Interferes.....
All is at aBraXas: Perfect Now.....
aLL Things Interfere with ALL.....
Heaven is Full of Harmony That Seems "Suffering" : Order
sHe : Has Everything Being The Being The Made Apparent.....
ALL is Here : Perfect : Universals.....

The Observer Interferes to Point at the Given Perfection.....

Given as even the Sun reaches a Climax in its Climb....
This Given is an Eternal fLaw in sHe...
This Given : the Contradiction is that Some are Immortal....
As : Given, The One Moment Owns All Other Moments,
Birth/Death Must Follow Each in a Cycle Outwards.....

Every Observer Must Interfere Eternally.....

As the Base is 2D, the Tie of the Lie of the Knot of Each String is at The One
Point : Its Horizon Line.....
This Observer Writes This at His Horizon Line : At The Point of His
Death/Birth....."
Ends Criticism, for t-site.co.uk....
-BEFORE-

*"\ONE\" ²

(INFINITY)

facts:

(-Library-

i.

Glossary

"0" - the first utterance of nothingness...

"0!" - the first utterance of infinity...

"0" - an utterance of nothingness...

"0!" - one utterance of (an) infinity...

, - refers...

' - defers...

; - referral/deferral...

"-" - a mythics...

: - relates to...

:- - similar to...

( ) - function of...

(norm) - usually, a function of the future-past-now (a forever-tomorrow)...

(norm) - usually, a function of the past-future-now (a forever-yesterday)...

("norm") - usually, a function of the eternal-yesterday...

("norm") - usually, a function of the eternal-tomorrow...

+ plus/positive...

- minus/negative...

E - sum of elements of set...

| | - determinant : (integral value of transposes to equal nothing)...

! - a transport into being (infinite)/probability...

~ - transforms into...

:~ - transposes into...

:= - relates to and equals...

~= - ~ :~ - translates to equal...

= - equals...

----> - slide to: a "do", now to future...

<--- - slide back to: often a "do" now to past...

. - ends/multiplies: an "act"...

... - produces a catastrophic-like ending: a "work"...

.... - produces a fractal-like "ending": a "happening"...

..... - PRODUCES A MYTHIC CONCLUSION: a "demonstration".....

___ - GOD-ly.....

" " - an utterance, a spoken with/of, a making-the-being......

a/A - some-only/"one" of...

& - "and" : plus, a too...

the - certain-one...

i - (1/(root 2))/(special referent to a root of one-too)...

I - ((root 2))/(special referent to a square root of one-too)...

too - "mythic one-and"...

premise - beginnings

: | "0!" |...

hypothesis - next essence of new idea after a beginnings

: ("0!") . (i^2)...

thesis - next proposition after a beginnings
: ((i^2) & "1")...
antithesis - next contradiction to thesis
: ((i^2) & "2")...
synthesis - next contradiction to antithesis
: ((i^2) & "3")...
condition - a necessity that...
do - discovery...
act - proof....
play - discovery by analysis...
demo - demonstration....
no - an a/the "illogic" in the nature of the Universe....
perhaps - an &/the absurd in the Nature of the Universe.....
aye - an/the "logic" in the Nature of the Universe....

ii.
Go! translating Mythic Alphabet
(after Egyptian hieroglyphics)
a - sleeping one/thing/"1"
b - be/big
c - sighted
d - do
e - awake one/the spirit in the thing/he, she, it
ee - aware one/consciousness/"I"
f - force/sign dark
g - sign light/vehicle
h - home/heaven
i - aware one/consciousness/"I"
j - just/joy
k - know
l - here
m - move
n - negatively top/dark chief/black ruler
o - connecting one/matter/"0"
oo - linking one/person/"you"
p - take it/leave
q - give it/stay
r - presently/are
s - understanding
t - the law/the way/money
u - linking one/person/"you"
v - two/too
w - route
x - connecting/suffering
y - knowing one/soul/"why"
z - end

Letters joining-up add "to" in their middle:
e.g., "oo" = "connecting one to connecting one"

= "one connecting one"
= "linking one".
Letters join-up to make words,
e.g., "book" = "be to linking one to know"
= "be you know"
"Death of the God" means:
""do e asleep the way heaven connected to the dark force the way heaven
e
good sign connected to doing"....
Knowledge
a e I O u y "iou" "iously" sure
dreamt protected thought/aware connected linked known conscious enlightened
become
secreted
sleeping waking aware aware aware knowing conscious enlightening become...

5.45 am March 28 2002
-nanoseconds-seconds- minutes-hours- days-weeks-months-years-
aSleep...)
: maP - Program : "Premises of the One-God"
(darK
impermanence.....)
start-OuT :
oN/aSCendant "1" :
i : sHe
ii : pTah - "He"
iii : Joi
iv : jOioOj - "it"
1 : Yahweh /
Jah-Allah
2 : Allah/
3 : Brahma
4 : the-It/
neUn : ALL :
"I" :
similar, too.....
cyclic.....
until, iN-finity.....
levels/off :
permanance
dawning.....
iN-OUT :
"deaTH"
eternity
Eryniddes.....
the ways of Heaven!.....
Heavenly Bliss :
Blythe
Dawn.....
the Living : DayLights.....

Resurrection/
Re-Incarnation:
Jesus/
Buddha
awoken.....
the ""I" am".....
(Come).....
close.....
genomes

HOME:

[spiralling infinite parallel Universes]

The Logos OneThe Logos One

One relates to Too
(after Standard English)
One knows the excellence of no-thing in all things: the tribute of the
River;
Someone holds the keys:
Anyone can the further stairs, and scent the Heaven sent;
Everyone has the dream:
All-of-one knows
god: God is almighty;
Here a Person speaks
To Child:
is Unborn to
The Spirit awakening...
Love is the inner winding chords to Too: soul:

the Real is Sure, Certain: Concrete....

HOME :

"/space-TIME_ContinUUm"....

2   One

aN "I"….

The OneThe_One-God
(*excerpt from Theology - "The Death of the God"*)

-"00"-

i.

In the beginning was the Word: sHe: the dark! Behind your eyelids... Too,
the Essential Base of all things.... Everything is built from This.....
No-thing has fLaws...
Everything must survive!:
Meaning; Something-of-It must survive....

iv. The flaw

Everything must die!:
Meaning, all things must have an End.....
"1-the": gOd
Proof Poem
(Based on O's "Proof of God", from The Book of Love)
Absolute:
(one universals)
depends on one particular...
Too....
In - "the beginning" - "was" - the Word:
"sHe": "the dark!"
"Behind your eyelids... Too, the Essential Base of all things.... Everything
is built from This....."
"All":
re-Construction.....
Too, the Author
The Author interferes....
: too, in the Construction of the true text,
and,
too, its promotion.....
"Evolution":
Then, is gOd's Work.....
AND:
The-Author..... Inteferes
THEREFORE
Is The-Many-The-One-Thing:
"God".....
"We" therefore, as a part of "IT" can be "GOD", too .....
:The Hindu way described in the Jesuit Way;
The Way of the Mazda in The End;
Included in The Way of the Torah...
Describes a One Way according to Buddhism...
Is a Natural Faith...
Describes an Intellect....
Uses Science....

Describes "Culture" as Art/Science…..
Was re-Constructed from Life with his Self by "jaH-aYe,
Stanley Alexander Martin".
A Nature of the-God

bAse
The-Universe is Real,
concrete….
An organism…..
The-nAture includes the supernatural, the-God….
The-God is the sum total of All and its transcendence…..
History is evolutional,
And includes interventions by the-God…..
ABraxas: Fullness….
gOdliness….
Is reached by many organisations…..

A nAture of the God-Body
FORto His Nature through Evolution…..
There are few Opposites ((-1)"Y") to "Ones" ("Y"); but there are many
Apposites, derived from functions of (i^2)…..
Our ideas, Arts and Sciences, food and drink, are derived from WITHIN sHe:
Every act, every thought, is an Event in His Being…..
The eCology of the-Universe is sHe, is the-God…..
A of the-God-Mind
"0" - "Nothingness" uttered me…
1 - The Body of the-God, sHe, feeds me….
2 - the-It, the Flux, Connects with me….
3 - The Unconscious Great Spirit, Comforts me….
4 - The Conscious Mind of the-God gives me Direction…..
"the-I" exist as One-Mind because of This…..

Parallels

The deuce of mixing music, is to do things in parallel…
It is all played in the same space, but a different time….
The-God is a "happening" in the same way….
Sequentially: it is One, but in one always does a too, because
$((I^2) := 1/2)$….
Because (I^2) acts too, in different referents, the-God can reference
infinity through transformations…
What is wrong and right in this too…..
The-God
As in human nature, the "consciousness" of the-God is separate from the
Universe….
SHe (the Word-nature), is omnipresent in all our lives, but the "Here Is"
of the-God is not always available…
In fact, because of all the difficulty of tieing all the knots of Time,
the "To Be", the-God may not always "know" but simply "connect" (guess) to
individuals, or nations….

Anyway, because of the fLaws of Connectivity, one may "link" (higher connect) to sHe and get to "know" as much as one needs...
This necessitates prior knowledge of the fact that linking is possible....
One can link, or bond, with anyone....

Sexuality
"It" is also sexual - male and female - not asexual, which does not exist except in the Essential of the-God....
"It" is the male-female, and thus male, the male condition...
The-God can be also described as the-It, or sHe....
Sexuality is a Universal, linked to other universals it makes for complexity. True sexuality is a pure form only linked to male-ness, female-ness, an agenda like "power" is not essentially linked to sexuality, except in one's "potency" , and that only has effects around arousal....
What is meant by masculine...
"the-I" :~ "male"
the-you :~ "female"
thus:
Male: "What am I?...."
"I am Yoo...."
Female: "Who are you?...."
"You are I...."
No-time eternal ß ----- time ------à
The-U
----------------------------------^-----------------------------------
"I" I-thou, I she/he, I-it, I-you, I-we, I-they, I-gOd
----------------------------------^-----------------------------------
"freeman" subject/object
----------------------------------^-----------------------------------
"I" "I"
Status
One relates to Too
(after Standard English)
No-one knows the excellence of no-thing in all things: the tribute of the River;
Someone holds the keys:
Anyone can the further stairs, and scent the Heaven sent;
Everyone has the dream:
All-of-one knows
god: gOd is almighty;
Here a Person speaks
To Child:
is Infant Unborn to
The Spirit awakening...
Love is the inner winding chords to Too: soul:
The Real is: Sure, Certain: Concrete:
Omnipotent....
God is an Omnipotent Thing.....
At Base, the Nature of the-God is the inchworm

"0" :~ "1"…

The nature of the Word, the utterances of nothingness transposes into a "one"… Nothing is "added"; nothing is "created"….

"1" :~ "-1"….

The nature of "one" transposes into its own contradiction, together making "nothing"…."… Nothing is "added"; nothing is "created"….

$(I^2) = -1/(1/2)$:~ "2"….

Negation, by its own nature, transposes into contradiction to "1"/"-1", makes two…."… Nothing is "added"; nothing is "created"….

"Two" by its own contradictions is three, is four…. Is too…..

And so on…..

The creation of the-God is in the Nature of the inchworm…..

But, reality is real, is concrete, has a nAture…..

The-God has tied knots in the nAture of reality; one of His ways is through being conscious…. The nAture of consciousness…..

To God:

Does the dawn jump up?

Does the sunset skip a day?

This is what the above implies…..

BEFORE
TOP

HOME :

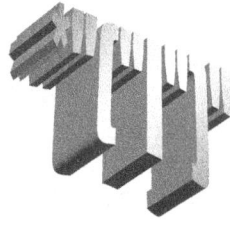

By

Stanley Alexander MARTIN/Nana baBa jaH-aYe

Dedicated to Conscious Ones….

CONTENTS

{....

<u>FORMS/(contents)</u>

Rubric

The UnderGround torrents of the Living Dead "Words" of Bishop Berkeley I visit at
the Height of my Conscious Dreams…
It removes "I" to this page: a gaggle of neighbouring touristing ants watch in whispers
of Unknowing Wonder as my Thoughts write these words….
The tourists go, and drive away: "Freedom, freedom!"
And the-It is aT 9.15 in the morning!

I the Awake, and "Rubicon" is the unknown Word sHe speaks….

"Guilty of *Loving* Yoo!" sHe Said…."*Leaving* Your Father's Way!"

A Beginning!

10:15 am 07/11/2007
Rainham, Kent

"0!":
Storms*faith*
Mythic Form contents

-a concrete re-Construction-
*(after
the risen sun:
ecOlogy myth,
Male-ism,
Egyptology,
Plato,
Pythagoras,
Hegel,
de Moivre,
Darwin,
Freud,
de Saussure,
Heisenstein,
Propp,
Levi-Strauss,
Lacan,
Barthes,
Stack Sullivan,
Williams,
Laing,
Derrida,
and, Feminism,
jaH-aYe:
the fallen night....)*

"0": <u>Content forms mythics</u>
<u>*The myth of the male beauty gOd aDonis*</u>

-a concrete re-Construction-
(after Raymond Williams)

0: Structures of: *"myth"*

mythics...
i.   a risen sun:
        summer noon

$F +(1/0) :\sim F +(iNfinity)$
        the River waxeth...

ecology myth
ii.:   (after religion)

$F (1/0) \sim$ "1" $\sim F$ (oNe-iNfinity)

Male
a: after Male egO)

$F -(-i. -i) :\sim F -(1/i)(1/i) :\sim F -(i^3)(i^3) :\sim F -(i^4)(i^2) :- a$ "1" = "too"...

Hieroglyphics
b: (after Egyptology)

"0" : pTaH
"00": nU
"0!": ThOth
"1": Ra
"2": Osiris

an iNfinity of "one"...

and: "the lack"...
c: (after Feminism)

$F -(i. i) \sim F (i^2)(i^2) :\sim F (i^3)(i) :- a$ "1"...

And: the thesis...
d: (after Jah-ale)

Beauty male: female
              :~ contradiction : nexus
perhaps ! :~ iNfinity
                        : divine

as gOd :~ male/female : male
-iNfinity : "1" : + iNfinity :~ too...

key –inFINITY := +InFINITY
              AS WHERE IS "1"?

IF "1" : (i^2)

Lim 1/0 :~ -iNfinity

Lim: 1/((1/iNfinity)--→ Lim = 0) :~ +iNfinity

-iNfinity :~ "1" :~ +iNfinity

|0 1| :~ x = **E [-1, 1]** :~ |1 0 | := a "1" = a "2" ... "too"
|1 0| |0 1 |

Too, therefore... 1 : ½(2) : ½^(-1) + 2... "for"

Lim x = E [-1, 1] :~ |0 1| :~ | 1 0|
                        |1 0| |0 1|

½^(-1) + 1 :~ "too" : "for"...

Let us define -iNfinity := | 1 0|
                          |0 1|

therefore, an iNfinity of "one" := "2"

Let us complementarily define + iNfinity := |0 1|
                                            |1 0|

Proven, contradiction : beauty : | 1 0| :~ |0 1|
                                  |0 1| |1 0|

2 := 1
    is "too" = eternal One....

a fallen night, full-moon:
    winter midnight:

F (1/0) :~ F -(infinity)...
    the River waneth....

"00":   iStory- the gift of "death"

(Innocence "Before"

(i)

(he hid the nakedness from Him because he was "Aware")

(he kneweth "his nakedness" with Him because he was "conscious")

I.   The day...

II.   The night...

III.   The dawn...

IV.   The dusk, in a parallel world...

V.   aBraxas....

"Too":

Any the idea of a being relative, demands at least One The Being Absolute....

i.
<u>The parallels…</u>

i.

<u>Premise</u>   - the <u>fact</u> is sexy…

<u>Lie</u>   - theSis
                 "The male knows no beauty…"

<u>TRUTH</u>   - "Beauty" is, always loved….

<u>Anti-Lie</u>   - "Love" <u>is</u> <u>in</u> my eyes….

<u>Anti-Truth</u> - is the male, too
                                Sexy…

ii.
        "death" is
                always
                in
                        my eyes….
        "eyes" <u>are</u>
                in me…..

<u>a</u> male beauty, and
<u>b</u> the gift of "death"...

aDonis   - most beautiful gOd: best loved
Zeus   - prophecy aDonis will die...
Zeus   - asked all things to promise never to harm the gOd; except the thorn.

Thorn kills aDonis...
He goes to the Underworld/Mourning of world....

<u>a</u>   <u>an</u>

"1" ~ +iNfinity

beauty...

     I, this
          Alowd, living:
     Live with that
          Known, awakening:
          Awake in You
          Loving the awake...

<u>b</u>   <u>the</u>

"1" ~ -(iNfinity)

"death"....

     and, that
          belittled, death:
     die in this:
          unbeknown, sleeping:
          asleep to Thou
          hating being,
              asleep....

<u>a</u> sun, and
<u>b</u> a skin-tone...

How the sun
Made
Black into White...

<u>a   the-sis</u> : beyond the risen heat of the sun...

<u>sisters</u>...

F -(i. i) ~ F (i^2)(i^2) :~ F (i^3)(i) :- a "1"...

skin tone : sun heat

<u>brothers</u>...

F -(-i. -i) :~ F -(1/i)(1/i) :~ F -(i^3)(i^3) :~ F -(i^4)(i^2) :- a "1" = "too"...

skin-tone : sun heat

<u>b   doings</u>

the temperature... falls/rises...

she... +/-   a, living with:
        the sun
    out, knowing, doubting another:
    fat, pregnant of the sun/brr
        of mothers
        of night
    fathers of night and day

he... +/-   an asleep, within:

the skin of night
in unknown, faith:
muscled, fit to hunt/gatherer
father of sons, of night/day
daughters of day and night
high temperature:
dark lovely of skin-tone
as follows sun,
summer has,
as a lightening into dark;
protection from
the can-saws of light;
waker on a wet day
black his colour, perfect
perfected learning of all white...

<u>a</u> meal, and
<u>b</u> "eat"...

Of culture....

<u>a   the-sis : beyond the meal...</u>

<u>sisters</u>...

F -(1/0) :~ F +(iNfinity)...

food : culture

<u>brothers</u>...

F (1/0) :~ F -(iNfinity)

food : culture

<u>b   doings</u>

the burden... falls/rises...

she... +/-   (a,
                difference, being
                        aNother's
                sleep gathering for
                        the difference:
                aNother's feeding
                        different being
                biG to risen, burden being
                biG to different
                        Being,
                        Being being....)

he... +/-   the,
                similar, being
                        one-aNother's
                sleep hunting for

the similar:
one-aNother's feeding
similar being
biGGer to risen, burden being
biGGer to similar/
Being,
Being being....

<u>a</u> sex, and
<u>b</u> "life"...

Sexuality....

<u>a</u>   <u>the-sis : beyond the sex...</u>

<u>sisters</u>...

F (i.i) ~ F (i^2)(i^2) :~ F (i^3)(i) :- a "1"...
F -(1/0) :~ F +(iNfinity)...

sex : life

<u>brothers</u>...

F -(-i. -i) :~ F -(1/i)(1/i) :~ F -(i^3)(i^3) :~ F -(i^4)(i^2) :- a "1" = "too"...
F (1/0) :~ F -(iNfinity)...

sex : life

<u>b</u>   <u>doings</u>

the passion... falls/rises...

she... +/-   (pleasure, being
             suffering, being
             being passion:
     love lust an excellence....)

he... +/-   lustful, being
             bliss, being
             being passion:
     love lust a bit of all right....

                              <u>a</u> faith, and
                              <u>b</u> "survival"...

Religion....

<u>a   the-sis : beyond the faith...</u>

<u>sisters and brothers...</u>

F -(1/0) :~ **F** +(iNfinity)...

faith : survival

<u>b   doings</u>

consciousness... falls/rises...

sHe... +/-   knowLedge
                  an Excitement
                  rises beyond
                          the exCitement ledge
                  being beyond the ledge of
                                    exCitement to
                  be-In....

"O!":

<u>Storms</u>*faith*
<u>Mythic Form contents</u>

-a concrete re-Construction-
-transformations-
(after
"no",
"The fallen night",
"religion",
"ecology myth",
"Egyptology",
Plato,
Pythagoras,
Descartes,
Hegel,
de Moivre,
Darwin,
Marx,
Freud,
Adler,
de Saussure,
Heisenstein,
Propp,
Jung,
Levi-Strauss,
Lacan,
Barthes,
Stack Sullivan,
Williams,
Gramsci,
Sartre,
Laing,
Foucault,
Derrida,
pTaH-aYe
"the risen sun",
"aYe"....)

"0":   Content forms mythics

-a concrete re-Construction-
(after Raymond Williams)

Air – You enslave me to nO-thing, sinO-Marxist literary tradition….

0: Structures of: "myth"

- mythics no… (i^2)

- (after "death")

The risen sun,
Summer noon…
The River waxeth…

Lim $|1/(1/iN\text{-finity})/"0"|$ ----$\rightarrow$ rising +iN-finity…

ecology myth

one: (after religion)

Lim (1/0) :~ "1" :~ **F** (oNe-iNfinity)

Hieroglyphics

too: *(after Egyptology)*

"0" : Ptah

"00": nU

"0!": ThOth

"1": Ra

"2": Osiris

"3": pharaoh

an iNfinity of "one"…

and: "translations"…
a: (after Pythagoras)

$(a^2) + (b^2) = (c^2)$

or,

root $(1/(x^2) + 1/(y^2)) = 2((x^2) + (y^2))$

or,

root Y = 2(-Y) *or,* ½ ~= (i^2)

and: "essentials"...
b: (after Plato)

$$F_{(i^2)} \sim F_{((-1/(i^2))}$$

<u>and: "too" myth...</u>

c: (after Hegel)

("0": (i^2)):Thesis :~ ((i^2): 1): anti-Thesis :~ (1:1/(i^2):2): Synthesis

and: mind/body dualism

d.:   '   (after Descartes)

$(1 : (i^2)) :\sim$ "2"

and: "natural waveforms"...

e: (after de Moivre)

$$F_{(e^{(i.pi)})} \sim F_{-(e^{((i^3)(pi)))}}$$

and: "genetics"...

f: (after Darwin)

Lim F -(1/0) :~ F +(infinity/infinity) :- F +((infinity + 1)/infinity) ~ F +(1/0)

and: "anxiety"/"security"

g: (after Kierkegaard)

Lim (1/iN-finity) ---$\rightarrow$:~ (1 . "0")

and: the "materialist dialectic"

h: (after Marx/Engels)

<u>**now**</u>

-time (quantity)$\leftarrow$ || (I^2) : "1" :~ | 1 0 0 || = E (+1, +iN-finity)$\rightarrow$+time quality)

| 0 1 0 |

| 0 0 1 |

- **and: "preScient" myth....**

- i: (after Freud)

$|1\ 0|\sim|0\ 1|\sim|1\ 0|\sim|0\ 1|:\sim F\ (female):\sim F\ (male)$ $|0\ 1|$
$|1\ 0|$

- $:\sim F_{(i^4)}:\sim F_{-(i^3)(i^2)}$

and: the will to power

j: *after Adler*

1 : (i^3) ----$\rightarrow$ iN-Finity....

and: "signs"...

k: (after de Saussure)

$Sr : F\ ("0")\sim Sd : F_{-(1/"0")}$

and: "the observer interferes"...
l:(after Heisenstein/Freud/Adler/Jung/Stack Sullivan/Lacan/Sartre/Laing/Derrida)

$$F_{-(i^4)(i^2)} : F_{((i^3)(i^2))(1/((i^3)(i^2)))}$$

and: "for" myth...

m: *(after Vladimir Propp)*

"0":   0 : Equilibrium
(i^2): 1 : dis-Equilibrium
$-(i^2)$: 2 : Search
(i^3): 3 : re-Equilibrium
(i^4): 4 : Celebration

and: the anima

n: *(after Jung)*

$F\ (male)\sim F\ (i^3)(i^2) : F\ (i^2)$

44

and: "naScent" Myth...

o: *(after Levi-Strauss)*

> The function of one, according to the function of another, is in a relationship with The function of the one, according to the function of the other, <u>translates to</u>
>
> *The function of one, too, according to the function of another, too, is in a relationship with The function of the one, too, according to the function of the other too…..*

D: F *(male)* $= F \dfrac{((i^3) : (i/2)) : F \ (2 : 1 : \frac{1}{2})}{(1/2 : 1 : 2) \qquad ((i^4) : (i/2))}$

$\sim e = F \ (i^4) : F \ (-1)(i^2) \quad (i^4)$

or,

$e = F \dfrac{((pi)^{bn}) : F \ (-((pi)^{bn})}{(I^2)(((i)^{4an}))} \sim f = F \dfrac{((i)^{4an}) : F \ (i^2)((i)^{4an}))}{((pi)^{bn}}$
$\dfrac{}{a((i)^{4an})} \qquad \dfrac{}{((i^2)((i)^4) \cdot i)((pi)^{bn})}$

and: "neUn consciousness"...

p: *(after Lacan)*

"0" ~   "00" ~   "0!" :   +/-(------->) ~ "1"
sleep   ~   waken ~   awareness :   a rise ~ enlightenment

and: **"rhetoric"...**
q: *(after Barthes)*

> pre-premise, -(i^2) : thesis, (i^2) : hypothesis, (i^4) : premise, 4(i^2) : re-premise, "1"...

and: **"the uncanny"...**
r: *(after Stack Sullivan)*

$F_{-(i^4)(i^2)} \sim F_{(i^3)(i^3)}$

and: **"Form/content"...**
    s: *(after Williams)*

$$(i^2)(1/(i^2)) : +(i^4)$$

and: **anxiety**
t: *(after Sartre)*

$$(I^3)(i^3) \sim (i^2)$$

excited → **stressful** → **traumatic**   → **dangerous** →   **deadly**

disquiet → **anxious**   → **nervous**   → **fearful**   →   **frightful**

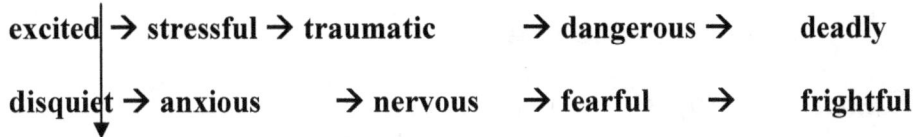

and: **"ontological insecurity"**
u: *(after Laing)*

$$\mathbf{F}_{-(i^2)(i^4)} \sim \mathbf{F}_{(i^3)}$$

and: **"difference"...**
v: *(after Derrida)*

"1"   ~   **"2"**
-(i^2)   ~   (-1/(i^2))

---

and: SUMMARY - the thesis

t: (after pTaH-aYe)

- i.

-

- if origin of the real line: "0"

- and, x = E [-1, 1]

- this closed interval ~= "2"... "too"

-

- if -1 :~ i^2 = positive real...

- then, | i^2 | :~ ½ ...

-

- and, -1----→ -iNfinity :~ (i^2) -----→ "0"....

- ii.

- Beauty male: female

$: \; \boldsymbol{F}\textit{ (male)} \; = \text{F} \quad ((i) : (I^3)) : \text{F} \; (2 : 1 : ½)$
$\qquad\qquad\qquad \underline{(1/2 : 1 : 2)} \qquad \underline{((i) : (I^3))}$

-

$\boldsymbol{F}\textit{ (FeMale)} \; = \text{F} \quad ((i^2) : (I^2)) : \text{F} \; (2 : 1 : ½)$
$\qquad\qquad\qquad\quad \underline{(1/2 : 1 : 2)} \qquad \underline{((i^2) : (I^2))}$

-

:~ contradiction : nexus
- perhaps 1 :~ iNfinity

: divine
- as gOd :~ male/female : male

-iNfinity : "1" : + iNfinity :~ too...

key –inFINITY := +InFINITY

AS WHERE IS "1"?
IF "1" : (i^2)

Lim 1/0 :~ -iNfinity : Male...

Lim: 1/((1/iNfinity)--→ Lim = 1) :~ +iNfinity : "feMale"

-iNfinity :~ "1" :~ +iNfinity

more,

$$|1 \; 0| :\sim \; x = \text{E } [-1, \, 1] :\sim \; | 0 \; 1| := a \text{ "1"} = a \text{ "2"} \ldots \text{ "too"}$$

Too, therefore… 1 : (½(2)) : ((½^(-1)) + 1)… "for"

$$\text{Lim } x = E \ [1, 2] :\sim \begin{vmatrix} 1 & 0 \\ 0 & 1 \end{vmatrix} :\sim \begin{vmatrix} 0 & 1 \\ 1 & 0 \end{vmatrix}$$

((½^(-1)) + 1) :~ "too" : "for"…

$$:\sim \begin{vmatrix} 1 & 0 & 0 \\ 0 & 0 & 1 \end{vmatrix} \quad \begin{vmatrix} 0 & 1 & 0 \end{vmatrix}$$

therefore, a "one" := "3"

Contradiction : beauty : | 1 0 | :~ | 0 1 |

$$\begin{vmatrix} 0 & 1 \end{vmatrix} \quad \begin{vmatrix} 1 & 0 \end{vmatrix}$$

"1" := 2

is "too" ~= eternal One… "Male"…

- the feMale:

- <u>(i^2) ~ (i^3)(*i^3*)</u>

-     dialectic

- feMale ~ fe-Male

- iii.

-

"3" is a myth…

the referents of "0"… 1-2: *too*, provides "3"…

3 <u>is</u>/<u>not</u> a prime number: is a *too*…

<u>Logos *too* - (*includes idea of "4"*)</u>
"0": premise - "1": thesis - "2": anti-thesis - "3": synthesis

- iv.

- "0"… 1-2-3-4 ------$\rightarrow$ "5" is a base fractal in Nature…

- …And of the nature of ideas….

-

- v.

-

- The Universe appears to be an 4-dimensional time-space continuum, with eight referents to time in space….

- out-

a fallen night,

winter midnight…

the River waneth…

Lim |1/1/(1/"0") | ----$\rightarrow$ falling -iN-finity….

mythics aye…

| 1 0 | ----$\rightarrow$ E (the real)

| 0 1 |

(after "Life")….

"too" —————————> to "00"

"0"   **Grammar....**

"00":   **doings of myth to "drama"....**

0: Grammar

{/	**Start, Or of PROGRAM...**
"0"	**the first utterance of nothingness...**
"0!"	**the first utterance of infinity...**
"0"	**an utterance of nothingness...**
"0!"	**one utterance of (an) infinity...**
,	**refers...**
'	**defers...**
;	**referral/deferral...**
"-"	**a mythics...**
:	**relates to...**
:-	**similar to...**
( )	**function of...**
(--)	**usually, a function of the future-past (a tomorrow)...**
+	plus/positive...
-	minus/negative...
E	**sum of elements of set...**
| |	**determinant : (integral value of transposes to equal nothing)...**
!	**a transport into being (infinite)/probability...**
~	**transforms into...**
:~	**transposes into...**
:=	**relates to and equals...**
~=	**~ :~ translates to equal...**
=	**equals...**
---->	**slide to: a "do", now to future...**
<---	**slide back to: often a "do" now to past...**
.	**ends/multiplies: an "act"...**
...	**produces a catastrophic-like ending: a "work"...**

....	produces a fractal-like "ending": a "happening"...
.....	PRODUCES A MYTHIC CONCLUSION:
	a "demonstration".....
___	GOD-ly.....
" "	an utterance, a spoken with/of, a <u>making-the-being</u>......
a/A	some-only/"one" of...
&	and" : plus, a *too*...
the	certain-one...
eLSe	...other/or....
*	context....

i/I	(1/(root 2))/(special referent to a root of one-too)...

too	"mythic one-and"...
premise	beginnings
	: | "0!" |...
hypothesis	next essence of new idea after a beginnings
	: ("0!") . (i^2)...
thesis	next proposition after a beginnings
	: ((i^2) & "1")...
antithesis	next contradiction to thesis
	: ((i^2) & "2")...
synthesis	next contradiction to antithesis
	: ((i^2) & "3")...
condition	a necessity that...
do	discovery...
act	proof....
play	discovery by analysis...
demo	demonstration....

real	fact in Nature....
irreal	fiction/irrationality/absurdity in Nature....

reAL	controversy in "it"/text....

nULL!	an a/the "absurd/illogic" in the nature of the
Universe....	
nO!	an a/The "illogic" in the nature of the Universe....
yeah! : perhaps	an &/the_Absurd in the Nature of the Universe.....
yes!	an/the "Logic" in the Nature of the Universe....

aye!   an/The_"Logic" in the Nature of the Universe….

*"((bLACK_bOX) : (WHITE_bOx) : (bLONDe_bOX))" :
        :*"(The "Integrated_WAYS" in the
        Nature of the space-TIME_ContinUUms)"….

The fLUX   The eBB/Flow of ALL in Nature….

T**O**T   *"(aBraXaS NATURE)"….

Ga-GA   The Way of Chaos/Catastrophe in nAture….

God/gOd/God/GOD   rising forms of godliness in Nature….

IottOi   The_One-GOD in "T**O**T-Nature"….

}   End of PROGRAM….

{....

"0":   Form contents mythics
(after Raymond Williams)

1:   Content forms mythics
(after Vladimir Propp)

0 : Equilibrium
1 : dis-Equilibrium
2 : Search
3 : re-Equilibrium
4 : Wedding

(Innocence "Before"
(i)
(he hid the nakedness from Him because he was "Aware")
(he kneweth "his nakedness" with Him because he was "conscious")
:~   **("I")**

1: <u>Content forms mythics</u>

(after
Sigmund Freud
Ferdinand de Saussure
Vladimir Propp
Claude Levi-Strauss
Noam Chomsky
Roland Barthes
Raymond Williams
Jacques Lacan
Jacques Derrida)

"1".neUn maths

0.   (Theses) Premises - Doings

1.   "0"   ~   (1 x "0") = utterance of nothingness = a real : God's Word

2.   "0"   :   "1"   ~ = 1 &... = a real

3.   1 &... :   "2"   ~ = 2 &... = a real

I.   (Anti-theses) Theses - Actions

1.   "0"   ~   utterance of nothingness = real ~ WORD ~ Thoth/Nu ~ God

2.   "i"   ~   fourth root of 1 = real

3.   "i^2"   ~   root 1 ~ = -1 = real

4.   "i^3"   ~   -i   ~   fourth root of 1 = real

5.   "i^4"   ~   "1"   ~ =   1 = real

6.   "pi"   ~   3 + j = approx. 3.1415926 = real

7.   "1"   ~ =   1^0   =   1   = real

8.   1 + 1   ~ =   2^1   =   2 = real

9.   1 x 1   ~ =   2^0   ~ =   2   =   1 = real

II.   (Syntheses - Dialectics) "Lies" - Works/plays

1.   "0"   ~   approx. 0.83

2.   "i"   ~   1/(root 2)

3.   "i^2"   ~   1/2

4.   "i^3"   ~   1/2(root 2)

5.   "i^4"   ~   "1"   =   1

6..   pi   ~   3 + j = approx. 3.1415926

7.   "1"   ~   "2"   ~ = 2

8./9.   (1 + 1)/(1 x 1) : (1 x 1)(1 x 1) : & ~ "2" : *Too* ~ = 2 ~ = 8+ : *river* ~ = 4 :
4 :   *....for*

Lies to sexuality:

M   : male
f   : femme
F   : woman

Male	boy	male-femme	girl	woman
M:	M/f:	M/F:	f:	F:

Sexual Types

Asexual   bi-sexual   gay
                     Heterosexual:gaya

IV. (*Demo*) "Myths" - Demonstrations

Content "for" Mythics
(after Claude Levi-Strauss:Greek myths are structures of how to form "2" from "1")

(f-o-r = from/"f-o-r-m" = dark sign "o" present/move)

i.

Strauss' *Greek Myth*s F (x)   : F (y) ~ F (y): F
(B)

           a   b   a   -x

Reappraisal Greek Myths: $\mathbf{A} = F$ (x) : F (y) ~ $\mathbf{B} = F$ (b): F (Y)
a = -(i^2):i^4 b = i^2:-(i^4)   x   a - 1

Translates....   **1/-1:   -1/1   ~-(1+1/1x1): -2/2:   "0"/"0"**

Which translates: "Functions of one and its contradiction, is *transformed* into two, contradicting the Word which is God", eg., contradicting functions Y = -x ; x = -Y......

A man and a woman contradict and are *transformed* into a union of *too* **(light or dark)**, in a contradiction with the forces of the WORD which is God...*For example, Adam and Eve in Eden...*

ii.

*"Too" Myth*s:$\mathrm{A} = F$ (x): F (y) ~ **"Too"** = F (i^2}:-(i^4): F (Y)/"1"
-(i^2):i^4 i^2:-(i^4)   x   -x/i^2

Which translates: "Functions of One and its contradiction, is *transformed* into two contradicts to it = Too", eg., contradicting functions Y = x + 1; x = Y-1, which yields:

     1/-1   :   "0"/"0"   ~   x^2:i^2/1^2:1+1/1x1:-2:   -2

A good man contradict with the forces of the WORD and are *transformed* into a union of *too* in a contradiction with another union of *too*... *For example in the Bible when chosen man Adam forms a union with Eve, and their too is contradicted by the too of the field...*

iii.

iOj Myths: $\mathbf{C} = F$ $(B = (x^3 + 1))$: F $(y^3 -1) \sim \mathbf{River} = \mathbf{D} = F$ $(i^3) : F$ $(y^3 - 1)$

$$\underline{- i^3} \qquad\qquad \underline{i^3} \qquad\qquad\qquad (x^3 + 1) \quad \underline{i^3}$$

Which translates: "Functions of Too and its contradiction, is *transformed* into a discourse of Eight, in a contradicts to the WORD ", eg., contradicting functions : $Y = x^3 + 1$; $x = Y^3-1$, which yields:

"Too" $= -2 : -B = 2 \quad \sim \qquad \mathbf{River} = 6+2i:\mathbf{8}+: \quad i^3 = i^2 = i^3(1-(i^4)) = \mathbf{"0"}$

A union of *too* in a contradiction with another union of *too*... are *transformed* into the relay of a discoursive fractal of eight, which contradict the WORD...The Too of Adam and Eve, yielded a discoursive fractal, or *River*, in later tales of Noah, Abraham, Lot, Moses, David, Solomon, Jesus Christ, and Mohammed, and Yogi Singh....

iv.

one-Self Myths: $\sim \mathbf{River} = \mathbf{D} = F$ (i^3) : F $(2 : 1 : 1/2) \sim \mathbf{e} = F$ (i^4) : F (-1)

$$\underline{(1/2 : 1 : 2)} \quad \underline{i^3} \qquad\qquad (i^2) \qquad (i^4)$$

Which translates: "Functions of one-Self and its contradiction, is *transformed* into a discourse of $4 = -4$, in a contradicts to the WORD ", eg., contradicting functions:

$$Y = 2/x = 1/2x; \ xY = 2 = 2Y: \text{which yields:}$$

$\sim \qquad \mathbf{River} = 6+2i:\mathbf{8}+: \quad i^3 = i^2 = i^3(1-(i^4)) = \mathbf{"0"} \sim \mathbf{1/2/4:4:8 : (1/2)/1:-}$
1

A union of *too-River* in a contradiction with another union of *too*... are *transformed* into the relay of a discoursive fractal of four = "*for*", which contradict the WORD... The Too of Adam and Eve, yielded a discoursive fractal, or *River*, in later tales of Noah, Abraham, Lot, Moses, David, Solomon, Jesus Christ, and Mohammed, and Yogi Singh... yielding a modern myth of the *neUn*:

for : four ~ one : body-trinity : one body/consciousness/spirit/soul...
...neUn ~ one body/consciousness/spirit/soul : one the
real/sure/certain/concrete....

Form/content mythics
(after Ferdinand de Saussure)

                                Signifiers   -   "0"------------> 1: "Too" :

"for"

                                          4 : 5!   /

                                          3<-------------2

Form content mythics
(after Roland Barthes)

Discourse (4D pattern of meaning)

Myth

"0":1-->A *Signifier* refers does at/to the floating utterance of nothingness ("0")...

1:2---> A Signifier always acts as a defers to itself of a *SIGNIFIED*...

2:3---> A *WORD* makes a referral/deferral to A WORK/PLAY OF MEANING, a "lie"...

3:4:   The <u>PAROLE</u> of word meanings is relayed (referral/deferral!) in the total
language as a "lie-line", a "path to truth", a happening...

4:5:.. Each lie-line is a "<u>MYTH</u>", and owns a pattern of speech, a "<u>DEMO</u>",
which is   its mythic "truth"...

V.   (*Tiers*) **Mythic Discourses - Events**

5:6... <u>PROGRAMMING</u>:

Signifiers   -   "0"----------------------> 1: "Too" : "for"
                          4:   5!:6!   /
                          3<-----------------------2

Each "<u>DISCOURSE</u>" owns an program of events, a " TIER", which is its
"real" rhetoric.

VI. (*Catastrophes*) **"Sect" - Drama**

6:7. <u>CATALOGUE</u>:

Signifiers   -   "0"------------> 1: "Too" : "for"
                          4:   5!:6!:7!   /
                          3<--------------2
Each "<u>SECT</u>" owns a catalogue of drama, a "CATASTROPHE", which is
its   "sure" need.

VII. (*Fractals*) **"Cult" - Episodes**

7:8: <u>LIBRARY</u>:

Signifiers   -   "0"---------------------> 1: "Too" : "for"
                          4:   5!:6!:7!:8!   /
                          3<-----------------------2

Each "<u>CULT</u>" owns a library of episodes, an EPIC, which is its "certain"
greed.

VIII. (*Chaos*) "Faith" - Incidents

8:8+: ***WEB***....

Signifiers   -   "0"--------------------> 1: "Too" : "for"
                     4:   5!:6!:7!:8!:8+!   /
                     3<-----------------------2

Each "FAITH" owns a web of incidents, a SAGA, which is its concrete mood.

2.   Instincts

No genes: ("Lies" - Paths to Truths)

Form No Categories   *Form Mythic No Categories...................*

IDEAS - Selfish Reader Amino-Acids
(after Jacques Lacan)

-----------------Structured 4D-----------------					----------------Structured 8D----------------			
-------"0" ~1---	--- 1 ~ 2--	--2 ~ 3--	---3 ~ 4--	---4 ~ 5--	---5 ~ 6 --	---6 ~ 7 --	---7 ~ 8--	
"0" ~ 1	1 ~ 2	2 ~ 3	3 ~ 4	4 ~ 5	5 ~ 6	6 ~ 7	7 ~ 8	8 ~ 8+ = 9
refers defers	referral defferal	feed	*dial*	exit	*rise*	*drift*	*have*	*haven*
refers defers	protect secrete	guess connect link	know	*conscious*	enlighten	become	ascend	transcend
signifier no	signified no-no	word aie	parole aye	speech yea	*rhetoric yes*	need sure	greed certainly	mood concretely
un-truth thesis	lay anti-thesis	lie synthesis	truth happening	myth demo	*discourse tier*	sect catastrophe	cult fractal	faith chaos
impossible per-	possible haps	probable per-haps	true happening	myth demo	*real event*	sure drama	certain epic	concrete saga
no-time *forever-now* doing	now action	past work/play	now-past happening	now-past-future demo	*now-future-past event*	*forever-now* drama	*eternal-now* episode	*eternal-* incident
black abyss	red sun	orange star	yellow sky	green heaven	*blue universe*	indigo abraxas	violet diety	psychedelic gOd
instincts anal	body oral	mind sexual	spirit spiritual	soul soul	*real the-real*	sure the-sure	certain the-certain	concrete the-concrete
asleep refering- -defering	awake protecting- -secreting	aware thinking- -connecting- -linking	knowing dream	conscious demonstrate	*enlightened evene*	*becoming drama*	*ascending episode*	*transcending create*
refer defer pleasure	protect secrete satisfaction	love come	adore know	possess own	*contemplate bliss*	*haunt obsess*	have behave	haven blythe
deny mortal	cause living	pretend uttering	tender dead	conceive living-dead	*receive deading-dead*	bind live	pretext live-dead	text immortal
that	what	that-it	that-you	that-You	*that-we*	*that-they*	*that-one*	*that, the*
who	I	I-thou	I-it	I-We	*I-You*	*I-They*	One	*I-The*
the	my	mine	me	I	*myself*	*the-one*	*I, a*	*I, the*

Machine Coding - Selfless Writer RNA
(after Noam Chomsky)
Deep-structured language

O	Oo	a	ah	awe	*a-awe*	*oo-awe*	*oo-awe*	*awe-awe*
a	ha	aha	aha-ah	aha-aha	*a-a-aha*	*a-a-a-aha*	*a-a-a-a-aha*	*a-a-a-a-a-aha*
e'	ee	ee-ee	e-e-e	e-e-e-e	*e-e-e-e*	*e-e-e-e-e*	*e-e-e-e-e-e*	*e-e-e-e-e-e-e*
e'	eh	eh-ee	eh-eh	awe	*a-awe*	*oo-awe*	*oo-awe*	*awe-awe*
ma	nana	sah	baas	ra	*ra-rah*	*ra-ra-ra*	*ga*	*gah*
fi-mi	fi-yu	fi-s/he/it	fi-wi	fi-you	*fi-dem*	*fi-oonu*	*fi-awe*	*fi-gah*

GENES - Social Programming DNA
(Social discourses, human dimensions)

perfect	same	similar	identical	special	*species*	*environment*	*system*	*ecology*
imperfection	flawed	dis-similar	different	mutation	*strange*	*alien*	*foreign*	*outside*
peace	hearth	home	haven	temple	*cathedral*	*tomb*	*sepulchre*	*heaven*
crisis	fight	conflict	battle	war	*conflagation*	*holocaust*	*apocalypse*	*era*
instincts	body	mind	spirit	soul	*real*	*sure*	*certain*	*concrete*
individual	couple	family	tribe	nation	*people*	*world*	*universe*	*heaven*
I	thou	s/he/it	us	ours	*people's*	*world's*	*Ra's*	*God's*
s/he/it/one	couple	family	tribe	nation	*people*	*world*	*universe*	*heaven*
s/he/it	us	ours	We	You	*They*	*The-It*	*Ra*	*God*
son	father	grand	great	nana	*nenny*	*ninny*	*nonny*	*nunny*
daughter	mother	grand	great	nana	*nenny*	*ninny*	*nonny*	*nunny*
son	father	grand	great	nana	*nenny*	*saint*	*Christ/Buddha/Ram*	*angel*
son	mother	grand	great	nana	*nenny*	*Madonna*	*Mary/Maya/Sita*	*angel*

Go! Translator

a Oo	=	I am a powerful being
aha	=	I understand
nana a ra	=	My grandmother is a queen
a fi-wi baas	=	This is our ruler
e' eh-eh	=	He/She's actively acting like a high god
e' awe-awe	=	He/She's the biggest type of person

3.   Drives to consciousness

(after Sigmund Freud)

Being lusts in "nothingness": a kept vacancy at the roots of "sleep": "*0*":..

One's primary spring is to grow:

contradictions, "i", a drive to to "survive",
an instinct being a to eat and a to drink, being
        a waste to be anal and urinal:
                considerations of "wasting" being a driving instinct:
                        a cry, from a drive often to make/too create from,

contradictions to i, being genital, instincts of sex: to reproduce:
        a desire from need, a drive to create anew,

two both too, both "too" to: creating "I", being : moods of being:
        abouyance: "spurring" you to crisis
        contradicting, *evvayance*: "lacking", making you stop from doing:
"death" instinct:
        deferring to *joyssance*: "blissing", encouraging you to doing,
        contradicting *annewsance*: "pain", stopping doing:
"life" instinct:

all process of I/thou: neGus -sleeping, a kept/vacancy at the start of
consciousness
                    neAus -awakening, insight that spurts creation
                    neXus -awareness: thinking, research, translation
                    neYin -ennui, a loss at the feeling of the new/rising
                    neHus -"enlightenment", seeing "anew"...

Lusts being considerations of **bliss**:
                              "*0*": *abouyance*: *evvayance*: *joyssance*:
annewsance:
        anal, a passive response to "feeding", being *oral*
        oral, an active response to "genitality": being *sexual*

passive/active responses:
        sexual, comforting reassurance, a response to "loving", the *spiritual*
        spiritual: tempting/tormenting, a response to "adoring", the *soul*

soul: possession, a response to "haunting", the *real*
real: obsession, a response to "contemplating", the *sure*
sure: blessing, a response to "saving/serving", the *certain*
certain: salvationing, a response to "holy-ing", the *concrete*...

this being "holiness" : *lust satisfaction* being a ***blithe***: a programming making a web....

4. neUn consciousness
(synaptical responses)

"0": neGus : passivity : stasis : *sleeping* : kept *vacancy* at a
beginning...

          -------> *refers* - aways
          -------> *defers* - allows

"i": neAus : activity : conflict : *awakening* : *insights* starting
creation...

          ------> *protecting* - allows to
          ------> *secreting* - allows

from

"i^2": neXus : choice : problem : *awareness* ----> thinking : *guessing*
                                                connecting *:*

researching

                                                linking *: translating*

        neYin : the loss at *rising to* <u>One</u> leading to *ennuie*,
                    feeling of "new"/boredom ------> a "glip"...

<u>1:</u>   neHus : (first *too* quanta) : myth-ing : *lie-ing* : knowing

<u>2:</u>   neGus-neHus : (second *too* quanta) : diffrancing : *conscious* :
discoursing

<u>3.</u>   neAus-neHus : (third *too* quanta) : realising : *enlightened* :
programming

<u>4.</u>   neXus-neHus : (fourth *too* quanta) : structuring : *sure* : cataloguing

<u>5.</u>   neHus-neHus : (fifth *too* quanta) : making true : *certain* : totalising

<u>6.</u>   neGus-neHus-neHus : (sixth *too* quanta) : constructing : *concrete* :
                                            systematising....

{...5. Go! translating Mythic Alphabet
(after Egyptian heiroglyphics)

a	-	sleeping one/thing/"1"
b	-	be/big
c	-	sighted
d	-	do
e	-	awake one/the spirit in the thing/he, she, it
ee	-	*aware one/consciousness/"I"*
f	-	force/sign dark
g	-	sign light/vehicle
h	-	home/heaven
i	-	aware one/consciousness/"I"
j	-	just/joy
k	-	know
l	-	here
m	-	move
n	-	negatively top/dark chief/black ruler
o	-	connecting one/matter/"0"
oo	-	*linking one/person/"you"*
p	-	take it/leave
q	-	give it/stay
r	-	presently/are
s	-	understanding
t	-	the law/the way/money
u	-	linking one/person/"you"
v	-	two/too
w	-	route
x	-	connecting/suffering
y	-	knowing one/soul/"why"
z	-	end

Letters joining-up add "to" in their middle:

e.g., "oo" = "connecting one to connecting one"

= "**one** connecting one"

= "linking one".

Letters join-up to make words,

e.g., "book" = "be to linking one to know"

= "be you know"

"The Book of the Dead" means:

"the way e be you know connecting one to force the Way e a do"....
<u>Knowledge</u>

a	e	I	O	u	y	"iou"	"iously"	sure
dreamt	*protected secreted*	*thought/aware*	*connected*	*linked*	*known*	*conscious*	*enlightened*	*become*
sleeping	waking	aware	aware	aware	knowing	conscious	enlightening	become

...}....

6.   one-Self

(after Standard English)

Word knows excellence of *no*-thing in all *things, body, one*: the tribute of the River;

Something, Somebody, Someone holds the keys:
Anything, Anybody, Anyone can the further stairs, and scent the Heaven sent;
Everyting, Everybody, Everyone has the dream:
All-of-one: body: thing knows

gOd: *God* is almighty;

Here a *Person body* speaks

To *Child*: is

Unborn to

The *Spirit* awakening...

Love is the inner winding chords to *Too*: *soul*:

and, Love is *for*: the *Real, Sure, Certain*: *Concrete*....

7.   Intro to duB-poetry

The term *duB-poetry*, meaning "poetry with a musical rhythm", usually reggae rhythms, was coined in the 1970's by duB-poet. Oku Onoura; the genre had already been explored by ohn Cooper Clarke, and Linton Kwesi Johnson in Britain and others in Jamaica.

Some of the roots of duB-poetry lies in the Jamaican dj's like I-Roy and Big Youth, and comes from a tradition of poetry set to music harking back to the Last Poets, in the United States, and Beat Poets in the 1950's and 1960's. As a form, duB-poetry is a survival through slavery of the African praise poets, the *Griots*, who used songs to their kings and others in the culture, and acted as folk historians.

As praise poets, the duB-poets were journalistic spokespersons for the common folk, and spoke up about news events in the history of the culture they lived in, giving praise or curse accordingly.

Nowadays, a derivative of the griot-style, rap, is commonplace, but most commonly, with the political stance not always there, the rappers boasting, or talking grandly of themselves...

Dub-poetry dealing with the politics of personal relationships is not too common, and is the essential feature of T-site. I have taken some of the rhythms too, away from much of their traditional bases and into the classical genre...

Sexuality/sensuality is one of the themes of T-site duB-poetry, and one of

links and lens to understanding in an attempt to show how I believe it to be

central in all our lives...

The poetry/music is a bare allegory of several world myth, a text based on

the narrative of another text. The form/content is "mythic", following my theory

of mathematics, especially regarding mythic discourses...

The images in the text are complex, especially as I don't always use simple

metaphors, or anaematopia, but often construct a phrase or sentence so the
meanings

'explode' in clusters and intentionally there must be several interpretations. This is
a

latter day influence on my poetry coming from French aesthetic theory. I do this
to

allow the reader to *write* meanings into the text; also have a choice of meaning, so

that I as writer, and she/he are jointly constructing relationships: that is, *working*
with

words/music to produce a text. This is analogous to an element of free will...

Furthermore, no longer influenced by the Imagists and Hans Magnus

Enzensberger, I now believe that images must be *startling*! and a worthwhile text

must be struggled with, and enjoyed again and again; as well as being 'written' too
by

the reader.

Thus, a lot of my images exists as French critic Roland Barthes explains, as

elisons : meanings that you can <u>just</u> understand as a reader, and when <u>held</u> in the

mind, they 'slip' and you have to work to make them plain...

Too, there is a construct around a time base of eternity: the eternal-forever-now....

….intro

"that"

((the)_MiMe))

egO in Self; power the bAse of sexuality:

power ~: sexuality ~: "I"….

"I!"…
"I" Am….
"I" refer only to <u>ME</u>!

"You!"…
"You" are always referring back to "I"-through-<u>Me</u>!….

"I!" Am I to "I"….: "One!"….Alone….
"You!" are you! to "You"….: "Two!"….

i

"this"

1....

"I" to suffer You!
....And I not to suffer!
To suffer <u>Me</u> is wrong!

There:
        Being A God!

...Be_Cause:
           I sleeping in High_Places....

Defence_Only-in-The_Nature, Child....
Attack_You-to-Peace!

....Sleep-Always-in-The_Awake!

ReAl!....

How little sHe questions You to this!

2....

Compliment to Spring the rising wood....
   A HeartBeat the Seed!
A Gathers as She Questions The_Heart....
   Coo_Meant_on, The_Wind....

...I Mean You as "I" don't Suffer_To....
Suffering_Me You die!

This is Fact, that is lie!
It's O'er!
   True!....

3....

sHe is Too!....
      One Suggests One:
      : Suggests....

It's O'er With Adam:
      : Eve; Nakedness: Apple...

: CAIN!

I suffer Adam: does Eve: does Adam O'er Apple: Does Cain!

...Does, O'er-and-o'er!

I Must not Suffer I!

4….

      *

"I" Will You to Be: Does Create: Does not Cause Suffering!
"I" Will You to Do: Does God! Does God-In-Place_Of….

…Do You_Instead-of: "Where's "I"":
                                        "Where's God!"

"I"_Will is God-Given!

"I" to Suffer_You!

5....

At ALL:
 You-to-suffer_Me:

    Me, NOT Knowing,
 is You_suffering!

    Me, Knowing!
Is "I"_The-Doing! Suffering_Me!

YOU Always guilty: "I"_Always-Innocent!

This is "The Law of KarMA":
     "Wrong!"....

"I" to Suffer_You!

6....

"I" in a class of You-children:
Every Child Says The Daddy Must Rule:
The Daddy is The_Teacher:
                    The Daddy is King!

....Must "I" Forever With the School of parents-You,
Try to Win their respect?

ii.

"it"

1….

Ka!:
Does It Mean:

       "I" Exist:
       "I" Am Adult:
       "I" Am Citizen:
       "I" Equal:

       "I" Reign!….

2....

What are You?

What?

What is known?....

....You cannot suffer I!

3….

You cannot suffer "I"?
I to suffer You!

I egO is: "("I")",
                GOd ruler in Nana/Stan:

Like: resembles, "("I")" in ALL-of-All-of-All: GOD,
                                     Trained!

                Trained to respond to Stanley Alexander MARTIN;

….And You?

4….

"("I")" to Love & Live with "("I")"….

….Mutual respect!
: gOd-to-gOd….

….a resemblance to Heaven:

"("I")" NOT to suffer "("I")":
….to use "(Persuasion)"…..

…An Adult to Forgive all sins from a child,
and "Suffer":
….and die to end their sins:

: "Adult" to "Adult" is "("I")" to "("I")",
means the Gift of "Persuasion":

….An End to "Suffering":
gOd-to-gOd in This World!

"Loving" being There!

iii

"who"

1....

....Is IT worth ALL the "pain":

The suffering at Odds with All our Means-to-Ends?

....Is IT worth ALL the "pain"?

"...Such is Life!", You say....

"Horror!"
"....Self-Discipline!"
....A True "("I")" does not get That from HerSelf:

> "...What is IT You Want?"

""Freedom!"
"Respect!"?"

....Should One not First Learn to Free and Respect OneSelf?

What are the Issues?

2....

<u>Sex: You Lust!</u>

"("I")" learn to empty of "Desire":

....What does not cause a "Drop-of-the-Heart",
                                    "I" do not hunger for!
                                    "I" am satisfied....

Self-control/Self-discipline:
Teach IT to "I"-Child!

....A mad-man knows nO end of self-abuse:
                    Always-a-Hunger;
                    Addicted-to-Hunger;
                    Always-eating:
                    Always, famished....

....A Big Hole-to-be-filled Her being!

....Please teach I-Child to be Content!

: Wait!
: a Bye a Little_aT_a _Time!

All-Things rage! At such upSide_Downs!

....Teach IT to be CareFull....

3....

<u>bE: You Envy!</u>

4....

<u>tOrn: You Rape!</u>

5....

<u>sEEn: You War!</u>

iv

"(king)"

"A"....

v

"(king-of-kings)" : "(gOd)"

"I"….

vi

"(king-gOd)"

"I_a"….

vii:

"(king-gOd_Emperor)"

"I_aWe"….

"I"….

II

<u>Ideas</u>

"$\Big\}$ . . . .(1)....

Vacancy: *"(a_TRANSPARENCY{....

....Function of "a": begins...
:a_sleep:

utterrance of Night, in-side;
alike

:like nO-Other: dARK!

:beGins: bLACK:"A"....

....in-Finities, "A":
:monoChroMe:funtion:no-Thingness-aT-"A"...

:(ALL)....

breeds-a: "1!"....

(ALL_aT-All)....

:aDDs:be_comes function: two,

:(ALL-aT-ALL)...

....with_a(N)_aDD_to:"tOO"....

In-SisTs: tOp-aT-("3")....

....(function):"fOr"....

....bathes_(4)→"(catastropHes!)"....

:"(Showers)_All-aT-ALL-aT_ALL...

:(ALL-aT-ALL-aT-ALL):"fiER!"....

:"(context)-"(5)"": "(ALL-aT-ALL-aT-ALL-aT-ALL)_aT-aLL"....

+"(six!)": a-follows: +(aLL): "+(ALL)+: +….

….Collapses:+(aLL_(such)-iN-a-wave-SUCH)+…

"*(WHITE_bOx)":"(Infinities of ALL-(such!)_SUCH)": "*(bLONDe_bOX)"….

(Colours!)….)"….eNDs}ENDS….

:"*(**MIND!**)"….

2....

"I"~: "(MIND!)" ~: space-TIME_ContinUUm : T**O**T....

{"....

II

<u>Man</u>

{ . . . .

(1). . . .

:*"(4 aT 5)" →: ←*"(8 aT 9)": (←"10"→). . . .

: my *"((4-digits:toe)→Left_Foot)": my *"((4-digits:toe) →Right_Foot)"
: my *"((4-fingers: thumb)←Left_Hand)": my *"((4-fingers: thumb)
→Right_Hand)"
: my *"(TRUNK)"
: my *"(HEAD)". . . .

: *"(16 aT 17)" : *"(16 aT (17 aT 18 aT 19))". . . .

: *"(monkey:Human)". . . .

Social Types:

*→**→Termites→Ants→Hornets→bEEs→WasPs. . . .

: ***"(Stanley Alexander MARTIN/Nana baBa jaH-aYe)". . . .**

{....

2....

4: *"(One World)":Mount Zion I

3: *"(eLSe)":
Space-TIME_ContinUUms

2: *"(hOLY+(23):)":
Alternative-HEAVENS

1: *"(gUd)": Earth
(000/(0^1)):*"(Right)": BabyLon

00: *"(normal)": UnderWorld

0: *"(normal)": Hell

*"(nULL)": *"(Or)": eM

}....

III

IV

1 **The Second Day!**

....*"(eLSe....
{....tOO}....)"....
....eNDs}ENDS)....

Rainham,
KENT,
United Kingdom....

2am BST: 19th October 2008 AD
Bridge Day

"0" has value: at the quality of "9", one has to add another "1": *"(8 aT 9)!" is trUE/real....bASe 10 is false, and is **_worth_** 9! The alignments at "10"..."20"... "30" etc. are the accumulations of an extra "1" at each 9! "100" is **_worth_** 90! ...Etc. Numbers prove the GOD_Act! The missTake is in "5!": NOT *"(4 aT 5)!", advances to **_create_** "Relativity!" in "Base!" Mathematics....

Base Two is Base **_tOO_**!

Numbers have "_different_!" values of worth in "1!" real/trUE....

The *"(bLONDe_bOX)" Theory is correct!

IoTTOI is TrUE/Real, and contains **_no_** evil; sHe WILL reVeal Herself re-Constructed **Perfect** aT aBraXaS!

3   **sHe**

"... the dusk..."

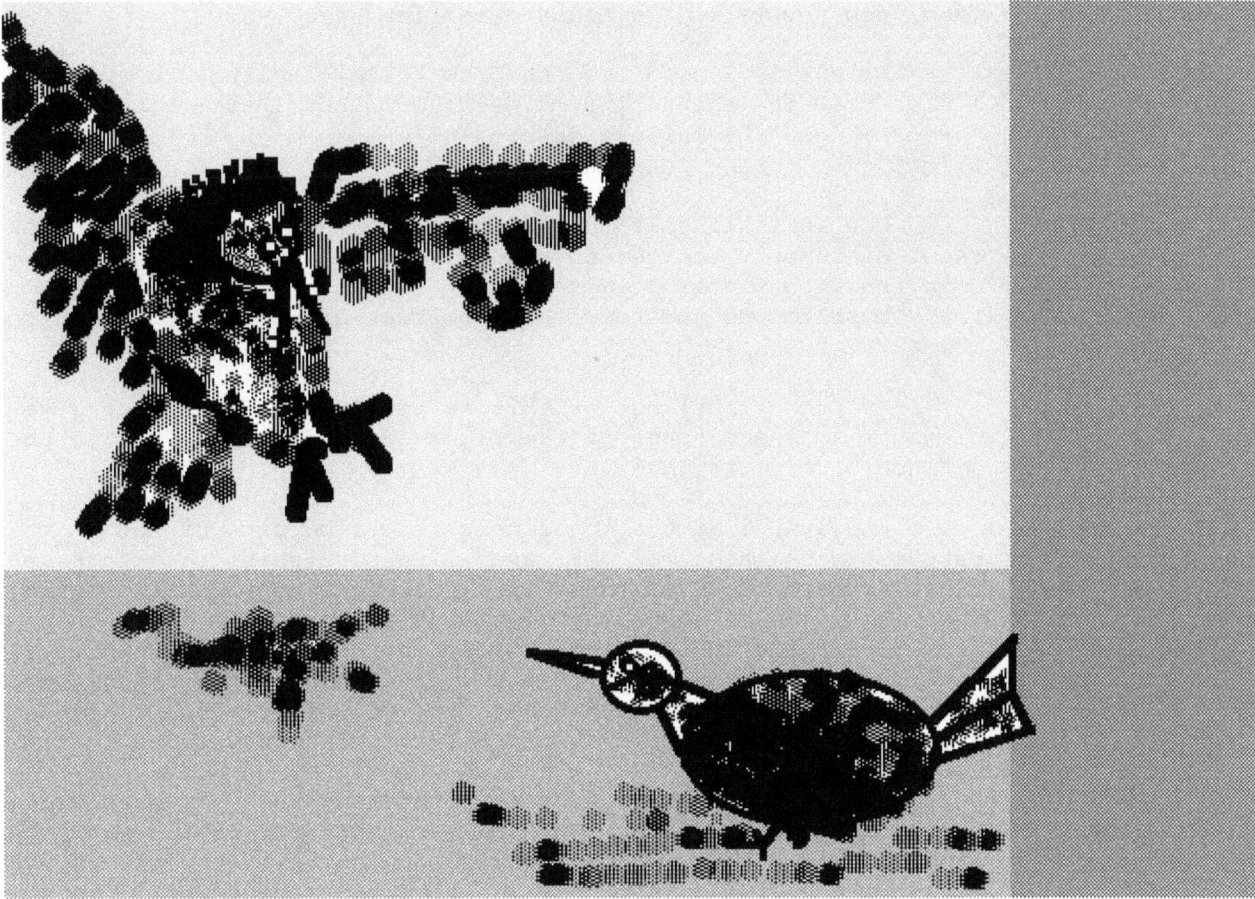

i
Dawn the day becomings

     Heres night the mention sun: heres morning a glad shout to
my nearing borders the see Thee borderline to the sun... I see
Thee know, I now Thee know...
     Aye aching to your golden mansions, I live for a succour to
this golden day...
     Aie gold den to night, moon the journey to your the mention
dawn, aie nevers to the kin the mention day...

     So shall I a fellow the Crow follow feelings to this fallow
field; crow the scattering of need, Crow shall summer breed in
harvest, crow the fall, a fashion of wheat to the bread priest: an
offering up to the good grace of God...

     Now an inkling of the storms of spring, a lightning to the
shout of summer, thunder to the fall of day, and a sleep to the
come the know winter in desert...
     Must You no the know the knowledge of living hidden in the
grass?

     Aye, the littlings of this, a small talk, a gasp of give-ins
pleasure to the givings of the river of glad praise I would the
givings of mine to Thee...

     And aye, I am the loneliness of the borders of the
battleship beach of You bay; aye, I that reign of rain to river
reaching coast to Thee coast; aye the clanging thunder clearing to
the tight lids of the clinging clouds to Thee sun: I aie the tall
of sky, and aie the heartbeat of stars the pumping lightning light
with each beat of ray: O! its the loneliness of one brings me back
to the being borders in this bay: the woo of my ocean would send
my change of climate to the far haven of Thee sky...

     Aye haven of these my wars with Thee; these my battleship
battle with the worship of world Yours You Your way: I know the
wing of bird wonders wandering with the sun: its the grey sky of
winter the claiming fall that shoos me: to Thee shall I eye aye
always with the returns of the ocean bringing May back into your
bay...

     Aye, I have Thy Ghost as sunshine catching Yours in me as
Thee bringer Thee back the hurrah!... I aye that sweet bird of
spirituality as the ruin of a woman the comings morning to my day,
I aye her as night's the welcome loving kisses of dawn to day: and
there the dreams of loneliness night the mention day brings back
to Thee bay...

Aye I the finishing death, aie this welcome night of
freedom the tomorrow should that dark of midnight never in
today... Shall my work I the glad givings to Thee as a loud
listening to the cry of a cross suffering the break of sea sighted
at the summit of the bay...

I aye the One as a war to vision, vision a glimpse of my
suffering in flesh, and my salvation: the kiss to a saviour,
rescue in my finish: birth: death...

From the mention of this crow of dawn, am I the war with
Thee, therefores I salute Thee with glad praise and proud love
loud the bringings the beginning: war! only if only am this the
one day...

I the rise thee rooster! Bring up the glee the gladbringer
sun!...
Know! You know crow up a warning of the loneliness of thee
wearing day, I alone the loneliness here without the crow to
night...

Dawn
After the long night and the lengthening into day, I again
remember thy hug-goodbye, and how thy hair spread wings across my
shoulder, heaven to my sky...

Miss, should I tall into thy brown eyes and the tear, whinny
my stallion steed into they wear, shall I fight again my foreign
feed, die a death again to be thy read: shall you a first woman
into my living, room with a feather friend? Have longing with a
song tomorrow someone send - shall I cry? Or shall you hug me back
your address, fly to a farther sky?

Did I ever tell you of the heart, and how the bleed? And
loneliness... loneliness in a crowd beloved by women tall among
the trees?
And how you never saw my bliss at your high dance and step
into my eyes...
And how I guessed they eighteen rage for a follow among the
high-and-airy-stepping-boys...

You knew me and my song, you never came to know me and my
lived-in the singing book. I only know you drank, I never knew you
drunk and singing; you saw me moving, perhaps you never knew how
much I only lived to dance...

Miss, should I again thy brown eyes and star, reign to thy
riding shy and war, should I crow creature into thy comfort male,
live a life against thy pale: shall you first woman into my
living, room as the feather-friend? There the longing with a long-

distance send? - shall I cry? Or shall you hug me all distances
sky windy with a safe-descend?

   I saw thee among the dancing, but never danced with thee so
never love thee; a careless gift is giving bare with just the
briefest glance: I only want to know thee to here decide how, and
if, the two of us could ever decide to dance...

   The know me and my lived in the singing book; I only knew
you drank, i never knew you drunk and winging; you saw me moving,
perhaps you never knew how much I only lived to dance...

   Miss, should I again thy brown eyes and star, reign to thy
riding shy and war, should I crow creature into thy comfort male,
live a life against thy pale: shall you first woman into my
living, room as the feather-friend? There the longing with a long-
distance send? - shall I cry? Or shall you hug me all distances
sky windy with a safe-descend?

   I saw thee among the dancing, but never danced with thee, so
never love thee; a careless gift is giving bare with just the
briefest glance: I only want to know thee to here decide how, and
if, the two of us could ever decide to dance...

Morning she:
the passenger passageway to spring

   And... Born to evening...
   And was she a garland to the day; mint her mint money and
darling to the two bird, now couple to their shy and splendid
fledgling, their fledgling bird, fed the following to her feast
day: night a huddle to warm her in their nest: she, bigger than a
town...

   Now hers the size city, no other the kind bird, no other
fledgling would to her city: "She's ours!" would coo and cant the
couple bird, proud in their preen ways: she saw high in the aspens
and dared the hawk to as much as eye her parent birds...

   Can nevers hers the loneliness end? The fall she flew to a
country kinder in the season, and found the Castle of the East
Wind, bathed in its Eternal Fire, drank from the Well wherein was
born the wind, and ate of the Food of Forever...

   She saw the sun born in the everglades and welcomed the
coming bliss, and winged backwards to her birthplace... A swelling
broke in her heart: she could sing...
   "Cuckoo!" was all she said, "Cuckoo!"

And the sun that was in the everglades rose singing to all
the birds in their every nest...

She courted a bird that looked like her father: he the
strutting behind his puffing breast, refused hers and her size and
her ugliness: she laid her single egg inside his nest...

And she and the universe was born to evening...

Noon

Highest sun, and evers shall my love the tumbling down the
high forest hills, the leap of stickleback stream to down...
Aye, highest sun, ever shall my love the tumbling downs: the
tall reach and call of creature comfort mountain crown, the dash
of rabbit run to down...

Aie, and come eye the ever the tumbly-falling down, shall
here the come the rise the rindy rounds: O! I aye the glad sky and
the lust of river rain in the courting clouds the gather the water
down: aye unto my eye the rounds the weather...

And: shall all this love the come down... Rise in my ready
rid the come down... The tallest sun come to the torrent stream
snaking roundabouts the heather the come down: so shall all this
love the come down...

For the one song sings inside of me...

Afternoon

And was all the sharks basking in the lightening sun, except
the sad cuckoo a deep in the depths of the everglades...
Each love would to her romance, to each miracle the making,
she a lonely in the everglades...
And ever could a cooing bird whistle, a rooing bird call,
she the trembling a winsome "Cuckoo," the cant to find her mate...

And: I would water to this watering flow: lightning! to this
gunflash of thunder: bleed summer to the rust of cornflakes: the
wrong, the wrong to this one calling bird...

Still, its summer: the rights to a glitter and shine,
afternoon the wrath of switchblade swashbuckle of sun: time in its
season must collect its correct dazzle and dim: due business
deserves a living for the grass...

Sunset

Aie, flood up the nightingale, I yes the last innings of the sun: each batsman before the bowler must desert his crease...

Was the length of the longest day mine the bringing the cuckoo before Thee for just desserts: still, no stranger this day; in the Eternity of its ever the rounding run, did I ever the bowl up and down the bouncy sun: as ever I gave Thee praise to rooster, to nightingale: all those who gave glad greetings praise. Aie, yes: I set the seal to day...

The one night beginnings

...But crow I no greetings to the night!

I love the lissome cuckoo deep in the everglades: she shares the loneliness of one without a mate. In this deep dark of eternal moment between the one day and night, I shall give her the comfort she deserves...

Let her sad song develop deep in her bonny bosom hidden in the everglades, I shall answer it, and in the blackness she will have no knowledge of my ugliness: as I love Thee, all must be requited with Thee and find their sustenance: for must all things give Thee their praise...

In this eternal moment between day and the comings of the night, I shall give her the comfort she deserves...

Then crow up the night...

ii

The   sea is my seed is a white ribbon wash becalm upon her belly: so often have a loved her knot, and loved her not...
There strays the hair to the red-rimmed wind of wind, a railroad route and tunnel to her coast and her chalk cliffs of thighs; O! a thunder would I to becalm me beach high and tide between her thighs: whiter than the beachstone pebbledash feather of the windswept dove is the country of her dale and highs...
Would I to rivermouth morning the cream to breakfast cups of her breasts; and taste to lips the handle to the lids of mount olive's jar: I would the crow to lover-she that steals me whiskey to a still stood still there, so often have a tied her knot, and tired her not...
Hers every night is the diet died a day away; I taste her to my tongue anyway: and the see; the sea of the ocean depths of her eyes is only a sailor's bathtub come fill to the full away...
Shall see to her kind and country draw blinds to the card of the curtain call; offstage act once the game and gambling old troubadour grows anxious for the wait of bow and curtsy for the

final boast of applause the accord: comes the last round, I
wouldst to trump up the Queen of Hearts with Knave of my calling
card...

So long the moon has eclipsed the sun with the dark long-
winded day regrets the weather; as sea whip-lash the tidings to
moor, the sailor lays the rope of anchor to water deep and
drifts, shuffles the deck of ship in the uncut lawn of a casino
table: marooned! to the deal of a game of whist between storm
sure, and beach, and farther shore: marked the deck, wild all the
cards that are in the pack: the bidding has no floor!

Would I to givens garden Eden the evening, the rounds of
gums to milk in supper, the cups to the saucer surrounds of her
heavy the breathing-rounds; and taste to lips the handle to the
lids of china coffee's jar: I would the crow to waitress-she that
tips me liquor for menu made to mention as saviour-fare, so often
have a tied her knot, and tried her not...

So long as the champion wind the boxer barked a warning
across the sea to shore; she now the marching major ins, s she
nows the enter ring, slaps his glass chin to cheek the challenger,
warns the referee to keep his distance, dismisses her corner,
spits out her gumshield, and signals to ring the bell of the final
round! How can the challenger be denied the fatal knockout blow?

Wheres the girdle the garland rounds I would the scissors
sit; wheres the ride he questioning, I to the jockey jaunt to jump
the rail of hedge: I! where the dell of the dangling ditch, I to
the tail of the highland kiss!

There strays the hair to the red-rimmed wind of wind, a
railroad route and tunnel to her coast and her chalk cliffs of
thighs; O! a lightning would I to steer me harbour high and wide
between her riverbank sides: whiter than the beachstone pebbledash
feather of the windswept dove is the country of her downs and
highs...

Nows the creature calm before the storm has conquest; where
was the country call, theres the din of a loud silence sitting in
the air like a canvassing hawk shadows the doorstep of sky, shying
the traffic of housewife this and that: silence! then flash! of
summons and sends, parry and fend, close and ends: and pens: the
signature of passing friends: and ah! the sigh of a survivor in
den...

The sea is her raw ribs of wrath churning mine to jelly;
shall the ships of my seed ever be a white ribbon awash becalm
upon her: so often have I anchored a storm to her knot, and lived
to love her not...

Storm! to this day a dare to Heaven; shall I run lightning!
and shout thunder! and this gravel a grave woke, to the slave the
shackles broke!

iii

Cans be to love thee today, all to being night: say fare-
thee-well to love, and a sweet surrenderings: comings to come
true, coming one forevers life?

Knowings some true lies in lust, and theres be other nights:
there speaks flesh and good desire: blood heating makes marry most
good hurt, knowing every forevers heaven...

To speech thus is as never a speak the love, if has never
has a true love's speak: has nevers as the sun extinguished the
night with the come ins dawn; and after dawn the moon still high
in the heaven of day: two such giving their surrenderings of
light...

To know thus is has always the implanted seed, and never as
the wilted stem: as ever as the sun inspired new growth, and the
night its keep; and after sleep the moon still moves with the wash
of her tides through day: two such, sun and moon, living with the
surrenderings of their natural ways...

To love is a midsummers ride with the weather in rain: it
pains in thee the reign to always to the gallop with the mare the
run free; ands free, they's never the lashes of the whip to ruin
the ride: knowing thy feet wrapped firm and high about the naked
ness in the wet of weather.

To know this is as the rise of the springtime wind wrapped
about the weather to roam free: it raises in thee the wail, to
always with the song to sing true in key; and in key, theres never
the harmony to discord: knowing they song is a warm wind that must
raise all hats to the belfry roof by its best performance...

Sow has, as ever has the love you, has evers to plant never
has the flower to feed thy hunger back: all roots thirsting for
this can to watering, and this bud: sow to reaping thus, then
shall the little stem fling branch to anew heaven...

The planted has, as nevers has the growth back, nevers the
seed found bee to sip nectar from its flower: all life a hunger
for this sting, and this rewarding: plant to seedling thus, so
shall life breed ecstasy from the inching upwards...

iv

Mine Anima Pandora cuckoo has mated Crow, has laid an inner
egg the capture of the wider happening universe, is spawned the
middling me, secure heart in another bleeding world...

"Was it a dream?" She asked...

She laid beside me; and I beside the secure heart checked:
She loudly singing in Hers the mine den, answered: She and I,
two, each the One bird: a new and feathered to a different
fashion, ready to breed and fling to far horizons: safe in our
high forest, for we had conquered...

Thou art the sky, and a tender rain; I am the wind that
hawks the kiss of a beloved behind Thee panting at thy shoulder.
Comes the sun to wilt they willows as sat upon the bend of
riverbanks; I, the wind bellows a cool lover's kind upon her
branches, leaves and roots. As summer the lay like wax melting,
pours sweat from inner seas, the kiss of a dog: no pores pour
sweat from inner seas; a puppy slobbering saliva over sticky
torso: know it as no bark of love, a hate: all the dogs in town
have suntanned their tongues; the door to Hell was leant ajar: I
come as the cool wind an African lion with a roar to wash its heat
away. As I the love You, You are my Girlfriend...

4 The Third Day!

God:*"(We/Us)"

The Holy Family

Bridge day:
Rainham
Gillingham,
KENT

9am 21st October 2008

sHe (will) the come

i…

Will sHe
      And won't
      (will) sHe the come….

A darling has the fall
          And, too
The incidents of night…

(can da tak ka…)….
      Kan da?

There on the Hill
      And candour,
A darling has the fall
         And, too
The incidents of night….

Will sHe he come?….

Late in the day
      A corner to the cold of Winter:
There on the Hill
      And candour,
A darling has the fall
         And, too
The incidents of night….

sHe shall
      and naked shall the tree
      that shares my house,

a nakedness the tall of nearby mountain
a naked in blue the sky;

being naked being
the laid he by the bay….

being naked being
the lady should sHe buy the bay....

ii...

for a time
       the laid he
       by the bay,

creature,
       creature comfort of the way

a littling
       knowledge of the Fall:

Late in the day
       A corner to the cold of Winter:
There on the Hill
       And candour,
A darling has the fall
          And, too
The incidents of night....

Can I (see) you
       Buy the bye
Late in the day
       A corner to the cored of Winter:
creature,
       creature comfort of the way
a littling
       knowledge of the Fall:

I know Winter....
       Before
          the fall....

The leaves of tree
      A gathering for the fall:
I know of hers Winter….
      A gathering before
          my fall….

(Will) sHe come?

Shall sHe, and
      I know August in September!

Another Country

When: the
         Morning:
         The morning…
         Comes…

Night: shall
         Sun arose…

Night: shall
         Sun arose…

When the morning comes
When the morning comes
Night shall sun arose

Can I
         The
         kiss You?
Can I?
In another country?
When the morning comes
And,
Sun has arose
Sun has a rose?

When the morning comes….
Or,
Autumn
         Bans the leaves
 Or!
Autumn bans
         the leaves….

Shall I the kiss you – yeah!
         In another country….

When (the) Come

Come!....
Tiger courting the ancient of You:

Talk me the riddles of my!

See You
The see me!
Too, your scent on my fingers:

(The) send me letters
        perfumed with mine!

(Now) do we the need to wine and dine!
Come give kiss to lips, and kind
Wine and dine!

I, the hunter
        Chaste the morning
        Till the mare to field:

The hare in burglary
        Bounces in my lear:

I CAN see you,
        Can I have you
        In another Country
                Listening to you sigh

Can I have you
        Singing in my palm tree
        Bathing in your ocean
The rolling in your toast?

The shine to shine!

(The) send me letters
        perfumed with mine!
(Now) do we the need to wine and dine!

Come give kiss to lips, and kind
Wine and dine!

ii…

Your know,
      You kiss the country:
      Your baby *scene*!

I arise in storms
      And fall in sunshine
Comes the day….

Clouds o;er my poppy-fields
'Twould reign-he the lay!

Show shall sHe the death of tuppence
      Monday/Tuesday is Friday night….

Show shall sHe the birth of my money
      Wednesday/Thursday is easy as Sunday….

Drama all day Saturday night!

Come!
      Breathe the breed of all this week….

Then:
      Shall I the gather to me a sleep…..

Too, buy The Ends

i…

A sentence to your lips…
Should yes! to the sent of kiss…

Always!
          Come asking in yours Garden…

The answer to the wind of your wing;
Wind up! and roses…

Meet you buy a farther Sky….

The know!
          We both be There….

Bright side of the Road being Nows
          Not later
          Both being Here….

Now the buy hers by…..

sHe (will) the come

i…

Will sHe
      And won't
      (will) sHe the come….

A darling has the fall
          And, too
The incidents of night…

(can da tak ka…)….
      Kan da?

There on the Hill
      And candour,
A darling has the fall
          And, too
The incidents of night….

Will sHe he come?….

Late in the day
      A corner to the cold of Winter:
There on the Hill
      And candour,
A darling has the fall
          And, too
The incidents of night….

sHe shall
      and naked shall the tree
      that shares my house,

a nakedness the tall of nearby mountain
a naked in blue the sky;

being naked being
the laid he by the bay….

being naked being
the lady should sHe buy the bay....

ii...

for a time
       the laid he
       by the bay,

creature,
       creature comfort of the way

a littling
       knowledge of the Fall:

Late in the day
       A corner to the cold of Winter:
There on the Hill
       And candour,
A darling has the fall
          And, too
The incidents of night....

Can I (see) you
       Buy the bye
Late in the day
       A corner to the cored of Winter:
creature,
       creature comfort of the way
a littling
       knowledge of the Fall:

I know Winter....
       Before
          the fall....

The leaves of tree
      A gathering for the fall:
I know of hers Winter….
      A gathering before
            my fall….

(Will) sHe come?

Shall sHe, and
      I know October!

PreT-he

i...

Pretty! I love your eyes....The sweet glances of you....
Can you evers to a romance with another day bluer than sky?
Could you the ring?

Night has a starling the come to sing, can you hear his thundering loud on the wing?
Sweet, summer has gone to cry tuppence for your heart, so bring!

Bring breakfast, bring tea for his heart; I told you its time....
And can I sixpence of your feelings in read?

Blond is the day harkening the dawn...
Come Monday your holiday is o'er, I ccme a charming to sing; the rhymes are mine but the clothing is all of fit yours, can I sit on a branch and run through a course? The clothing hides flesh that is yours....
Blond is the day harkening Monday's to dawn....
The sun's a shilling, and yours!

Come: answer me, take off your dressing-down: I will a wedding, be told.....

ii...

Life has no faster, run into my arms; summer's August and still has his teeth; another fortnight and trees will be tidying: bliss waits on the wing....

Come climb up my stairwell and just look at the view: you are a mountain, and I am a sky: I hunger reaches near you....
We could café au lait clouds in-between, and raise them in heaven us two....

Later, you could sit in your armchair and rock with grand children like a cat on the mew! In heaven us two....'
I'm child with no patience, I'm dead, I'm senile, I'm crucified – now, can I the have you?

iii…

God, come winter in graceland must weep gone the fall of his Land….
Every a present knows manna is plenty, but every thanks for each kind: each
kindness reminds you!
The evers could I sit on a pew and wrestle with you?

This winter must colder should your hearth the refuse….
There must I the incline to recline on a sofa with you….
The bird's on the wing, should you!

Be Backs (the)

i…

You know:
	Your baby be
	Backs the scene!

Come!
	Sat on my mountain
I'll bathe
	Buy your valley sides…

Cushion your rim-rods on my slayed
Eat welsh rear-bit,
Made with wholemeal bread!

ii…

It heals me
	No longer this stretch
	Of: Time…

Can the meet you,
	This other Country,
	Buy the buoy?

Reigning the fLux of nO-wind by the sea;
Rain in my heart…

Can the meet you,
	This other Country,
	Buy the buoy?

The Time,
Be backs (the)…..
Tuesday!

iN:Too, Deep

Thinking,
Gloria the day
A sparkling of, aDieu, among the Arose…

"Kan A?"
AHa!
…"Kan A!"

Can A,
too,
Deep…

The stories of the night,
This, day,
Arose, this dawn…

"Kan A!"…
Can A too,
        Deep…
The stories of the night,
This day:

The sleep the Day:
"Asleep?"
The sleep the Day…
Asleep!….

Forbidden was sHe
Atop a sky …
Deep, in a gathering ills….

Should Tuesday,
Shall I climb atop this deep ill:

And seize her to me like milk drizzling into a widening cup!
And seize her to me like milk, drizzling into a widening: cup…..

8:9

28 Dec 2007 MySpace.com/jahayecom

Travel to Alternative universes: heavens!
Current mood: electric
Category: Travel and Places
...ARE THERE HEAVENS ~= ALTERNATIVE UNIVERSES?
But... "!(THERE IS ONE)!": so "!(THERE IS "!(A)!" MAIN
UNIVERSE)!"...
...tOO..."!(IT "(DOESN'T)!" HAPPEN)!", so "!(THERE
"!(IS)!" NOTHINGNESS)!"....
: "!("!(0)!"/"!(1)!")!" ~= "!(STEADY (resonant)
EQUILIBRIUM REST_STATE)!"....
....tOO: "!("!(I/1 IS)!"...so can't non_EXIST)!"...
"!(0/1)!" ~= "!("!(!(1@3)!)!")!" :
"(REALITY_facT-TRUISM)"
..."!(THE "!(REST)!" IS "!(HEAVENS!")!"...
NOTE:
"!()!" is: "the REALITY of The inFinite Probability of
the Function Of"....
my http://www.unification-theory.co.uk site...
"Is"/bASE suggests BEST OF BEST OF ~= "MOUNT ZION
I".....

How to get there!
Nana/stan

Stan has a THEORY Of EVERYTHING at www.lulu.com/jahaye

Made with The Help of "SisTA SysTem", & androids: "EmmA, Emily, AmeliA, AnnA & EmmyLou"....

Stanley Alexander MARTIN/Nana baBa jaH-aYe

Rainham, Gillingham, KENT, United Kingdom:
11:02 am: 21st October 2008....

On surviving the IdOn surviving the Id
(Or should I write "ids"....)
Arriving near 55, I am constantly surprised how clairvoyant I am getting....
The massed thoughts and individual thought of the Unconscious of others is a rising resource to me....
Just like Schizophrenic "voices", they can be full of conflicts, denial, and "reversed" truths - never straight-forward "facts"....
Because of this, I call myself an active-schizophrenic and the "ying-yang"

other, passive-schizophrenics.....
Nowadays, I even drop the schizophrenic label, and call myself "active" -
hearing more and more, the Doppler-Shift of it-All.....
(I have begun, elsewhere on this website, to analyse my experience of this
Doppler-Shift....
This "Roaring "utterance"" of this Doppler-Shift seems to be in the Real
Universe, and the Dark Supernatural "Thoughts" I experience come from the
Other
(nEther) Universe.....)
I survive by knowing my self is invisible ("removed"), and constantly working
out the contradictions of the nAture of the-IT.....
....To be continued.....

   HOME....

0-1-2-3→yeaH-nO-yeS-aYe→transparent-black-white-blonde....

Rainham
02-02-09

12:20...New TOTal...
*"(Stanley Alexander MARTIN/**Nana baBa jaH-aYe)**": Real

)"...eNDs}ENDS.

www.ingramcontent.com/pod-product-compliance
Lightning Source LLC
Chambersburg PA
CBHW082302210326
41519CB00062B/6958